# SOLID GEOMETRY

*Alan Hoffer*

**PATTERNS IN MATHEMATICS**

Editor: D. E. Mansfield

# Solid Geometry

P. A. CAINE

1972

**CHATTO & WINDUS**

LONDON

Published by
Chatto & Windus (Educational) Ltd
42 William IV Street
London WC2N 4DF

*

Clarke, Irwin & Co. Ltd
Toronto



ISBN 0 7010 0480 0



Printed in Great Britain by
Cox & Wyman Ltd
London, Fakenham and Reading

# CONTENTS

# INTRODUCTION

IN this book we take a look at some shapes which fill a space. These we call SOLIDS and the study of them is known as SOLID GEOMETRY.

If you look up the word solid in your dictionary you will find that it has more than one meaning. It can mean, 'not hollow' and also, 'a shape having three dimensions'.

A dimension is a measure of length.

A line has just one dimension. We can measure its length.

A flat shape which has no thickness can be drawn on paper. It has the two dimensions of length and breadth, and covers an area of flat space, or a plane.

The solids we are going to explore have three dimensions, length, breadth and height. To show three-dimensional shapes on a flat page we need the special sort of drawings you will find in this book.

The red and green spectacles are used to view the drawings which will then appear three-dimensional and stand up above the page.

Your teacher will tell you about the correct method of using the spectacles to achieve the best results.

# RECTANGULAR PRISMS

USE your red and green spectacles to study the drawings on the opposite page.

These sorts of three-dimensional shapes are known as RECTANGULAR PRISMS.

Can you see the three dimensions, length, breadth and height?

How many ways can you find in which these shapes are similar to one another?

Write them down.

Look again and this time note any differences you can see.

A rectangular prism has six faces, top face, bottom face and four side faces.

What can you say about the shapes of the faces of these prisms?

The faces meet along straight edges.

How many edges are there for each prism?

Each corner of a prism, where three edges and three faces meet, is called a VERTEX. The plural of this word is VERTICES.

How many vertices has each of these prisms?

Say which of these prisms A, B, or C you think is the largest.

One of these prisms you will probably recognise as a cube.

A cube is a special sort of rectangular prism.

Look at the one opposite and try to say in which ways a cube is unlike the other rectangular prisms.

What can you say about the edges and faces of this cube?

Sprinkle a few grains of salt on to a sheet of dark paper and examine them through a powerful magnifying glass.

What do you notice about the shape of each grain?

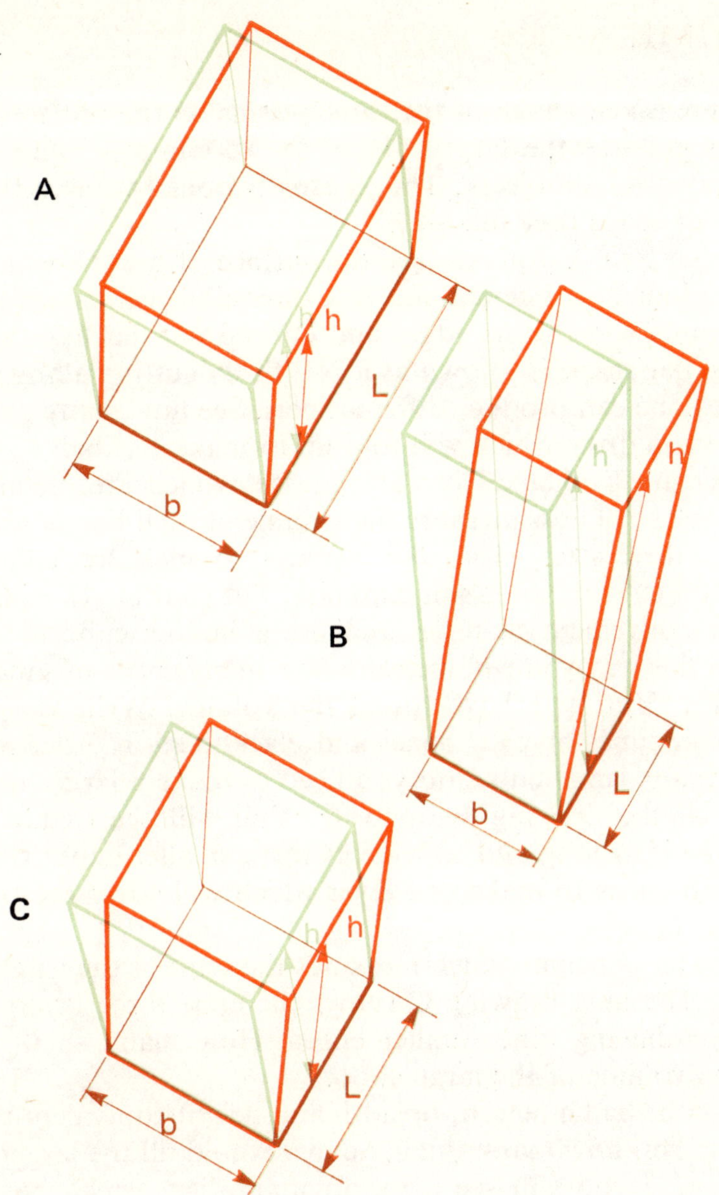
A
b
h
L
B
b
h
h
L
C
b
h
h
L

# VOLUME

You were asked which of the three prisms on the previous page you thought was the largest. How do we compare the sizes of three dimensional objects? This is usually done by measuring the amount of space they fill.

On page 5 in A you will see the surface of a cube which has been 'exploded' or opened out. If a three-dimensional shape can be cut along some of its edges and opened out flat like this, the resulting flat shape is known as a NET. By cutting along different edges one can produce different nets. See how many different nets you can draw which will fold up to make a cube.

Collect a set of boxes of various sizes which are rectangular prisms. We shall now measure the volume of each box in order to compare their sizes. The unit we use to measure volume is normally a cube of some standard size. Fill your boxes with sand or other convenient material and use a hollow cube to empty them. In this way you will measure how many cubes of sand each box holds. You will have found the volume. Write down the volumes in cubes of your boxes and place them in order of size.

How many small cubes do you need to make a larger one?

Look at the drawing B opposite. You will see that when a large cube is as long and as wide as three smaller cubes we need nine small cubes to make one layer which will cover the base of the larger cube.

A cube has a height which measures the same as the length and breadth. The next drawing C shows the cube three layers high, each layer having nine smaller cubes. How many small cubes make the volume of the large cube?

If the cube had a length, breadth and height equivalent to two smaller cubes, how many smaller cubes would fill the larger one?

On page 7 you will see a rectangular prism which has been opened out to make a net. Boxes like this are useful containers

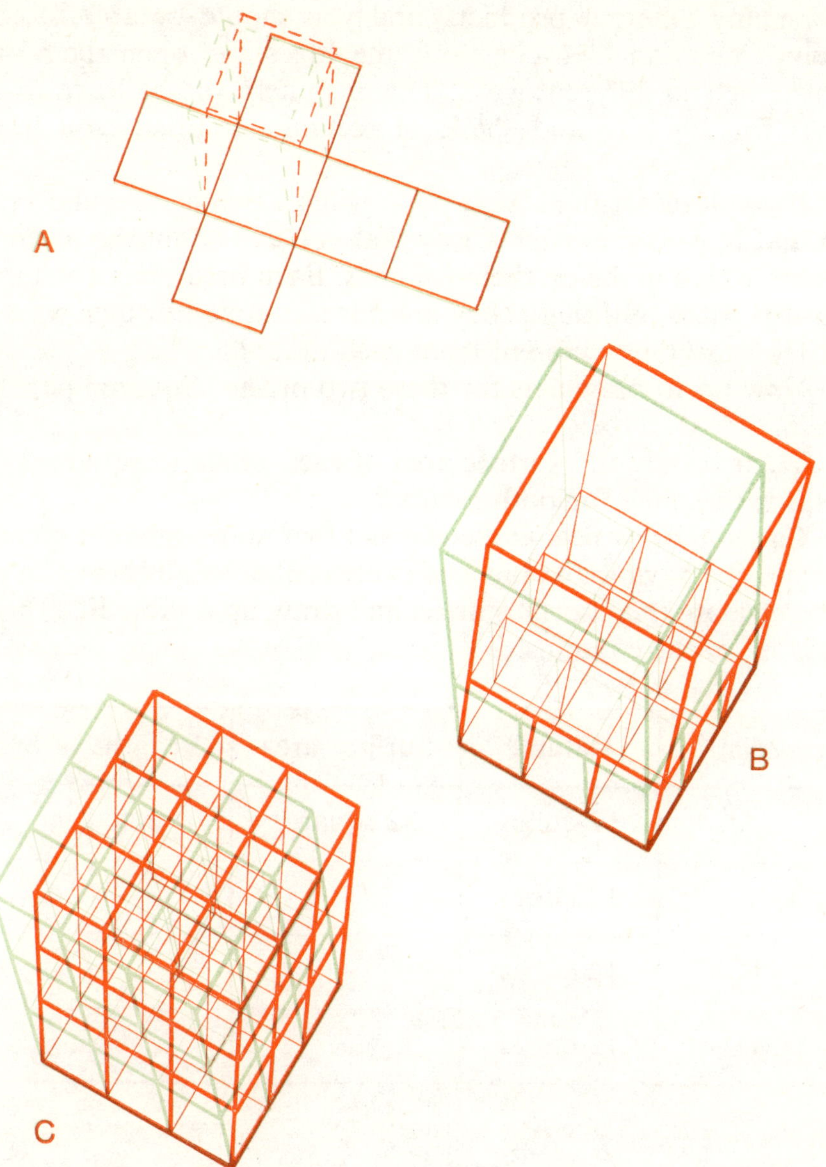


A
B
C

for many different products, and you should be able to collect some. You can then cut along the edges and open them out to make nets in different ways. This may help you to learn to draw nets for any size and shape of rectangular prism you wish to make.

If you look again at page 7 you will see two rectangular prisms, A and B, drawn in such a way that you can count the number of cubes which make up their volumes. Each prism has a volume of twelve cubes, although they are not the same in other ways.

How are they different from each other?

Now try to draw nets for these two prisms. Squared paper will help.

What is the total surface area of each prism in squares?

Is it the same for both prisms?

Can you draw nets and construct two more different rectangular prisms having a volume of twelve cubes? Call them C and D.

Calculate their surface areas and draw up a table like this one and fill in the columns.

| Prism | Volume | Surface area | Lengths of edges |
|---|---|---|---|
| A | 12 cubes | 32 squares | $2 \times 2 \times 3$ |
| B | 12 cubes | | |
| C | 12 cubes | | |
| D | 12 cubes | | |

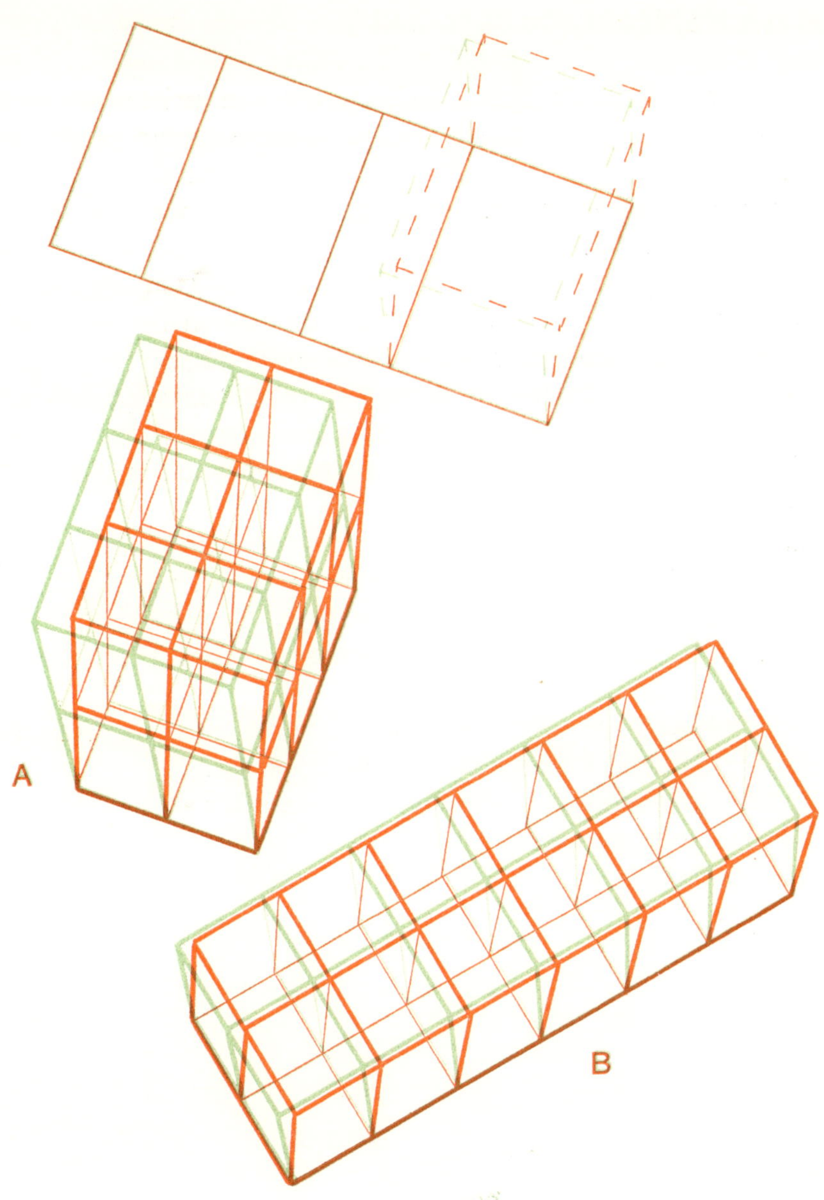
A
B

# SHEARING

USE your spectacles again to examine the prisms on page 9.

You see that the drawings show a rectangular prism which has been transformed by pushing it over to one side.

When a solid is transformed in this way we say that you SHEAR the solid.

This transformed shape is called an OBLIQUE prism.

We must ask whether there is any change in the volume and other measurements when we shear a solid shape.

Your teacher will show you how to take measurements of the lines on these shapes when viewing them through the spectacles.

Measure the edges of both shapes and say whether the shearing has produced any changes.

Examine the drawings again and write down the points of difference you notice between the right prism and the oblique prism.

Make a rectangular prism for yourself from Plasticine, clay or perhaps balsa-wood.

Decide how you could make one cut and re-arrange the pieces to form a sheared prism like the one opposite.

Is there any change in the surface area after shearing?

Has the volume changed?

How do you know?

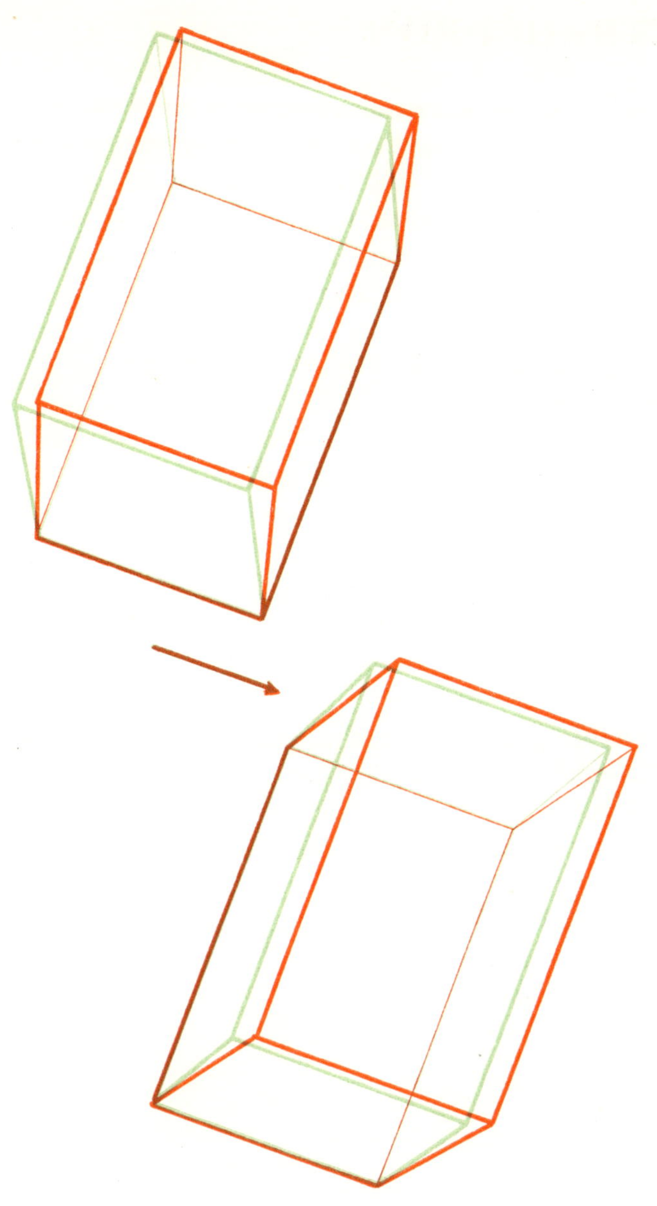

# THE TETRAHEDRON

ALL of the solids we have examined have been of the hexahedron family, that is to say they had six faces. However one member of the family, the cube, was different because all six faces were identical squares. We say that a shape is a REGULAR SOLID if all its faces are identical regular plane figures and its vertices are indistinguishable from one another.

Examine the shape opposite.

It is called a REGULAR TETRAHEDRON.

How many faces has this figure?

What shape are the faces?

How many edges are there? How many vertices?

The drawings show how the tetrahedron has been opened out to make a net.

Use this net to make your own tetrahedron.

Which way up should it stand?

Can you make a tetrahedron which is not a regular figure?

One advantage of these special drawings we have in this book is that they enable us to take measurements inside a solid. This is difficult to do with a solid model.

If you look again at the tetrahedron you will see a line rising vertically from a point in the base to the top of the shape.

Measure this line. It is called the vertical height.

You will see the value of these measurements later in the book.

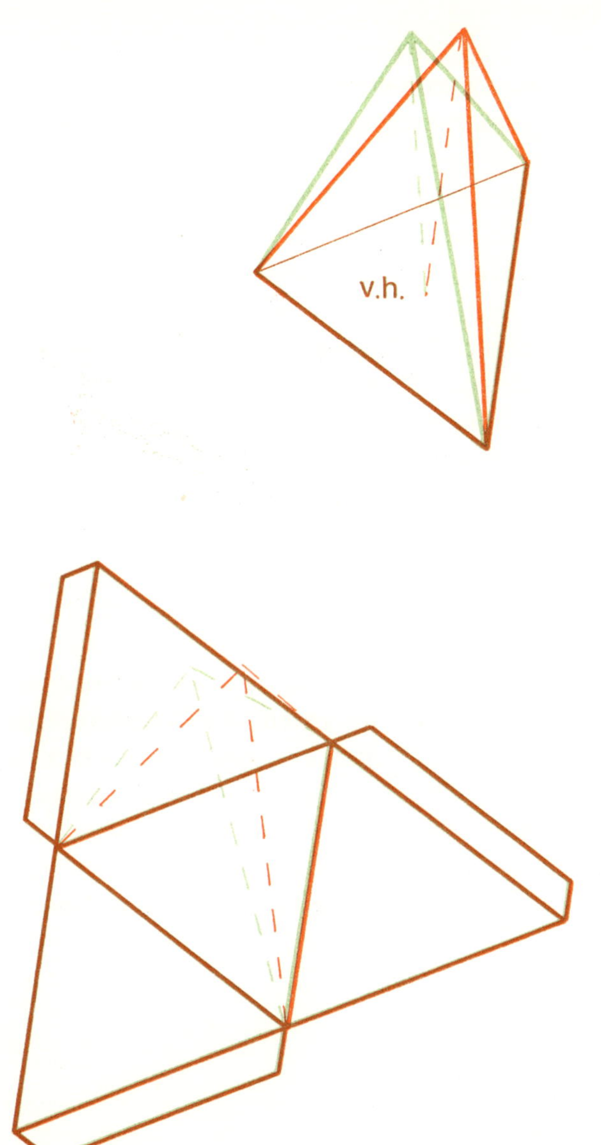
v.h.

# PYRAMIDS

ALMOST everyone has heard of the famous pyramids which have stood in the desert, near the River Nile in Egypt, for thousands of years.

The first drawing on page 13 shows this sort of pyramid. The regular tetrahedron is another member of the same set of solids.

Can you say why the tetrahedron is known as a triangular pyramid, and drawing A opposite a square pyramid?

How would you describe the pyramid shown in drawing B?

What difference do you see between the two pyramids?

Answer these questions for each pyramid.

How many faces has each shape?

What shape are the faces?

How many edges and vertices are there? (Notice that now more than three faces may meet at a vertex.)

These pyramids can be constructed from nets as with the one we used to make the tetrahedron. However as the number of sides increases it becomes more difficult to join them at the top of the pyramid. There is another method of construction shown in black on page 13. This method involves two pieces, the pyramid and its base. The pyramid is joined to the base by glueing the curved flaps on the bottom of each face.

If you have a transparent plastic protractor your teacher will show you how to use it to measure the angles on these drawings where the triangular faces meet the base.

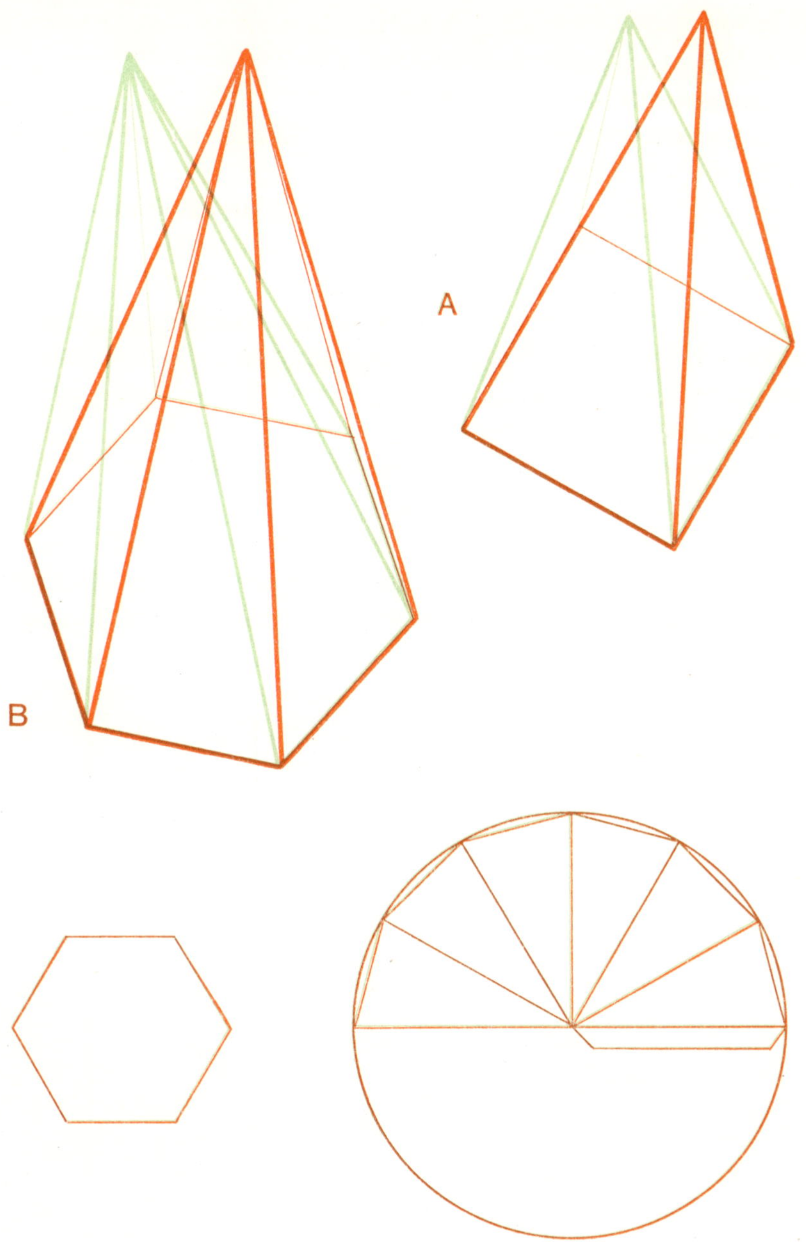
A
B

So far we have met shapes which belong to two important sets of solids. (a) The set of Prisms. (b) The set of Pyramids.

We shall say that the prism is a solid having the top and bottom faces identical and that these faces are joined by others which are rectangles or parallelograms.

Pyramids stand on a base from which triangular faces rise to meet a point called the **APEX**.

Study the two drawings on page 15.

Can you say which of these is a prism and which is a pyramid? Or are they both prisms? Or both pyramids?

How many faces has each shape?

What shapes are the faces?

How many edges and vertices has each shape?

How many faces meet the vertices?

Are these shapes regular solids? How do you know?

Now we shall collect together all the facts we have discovered by studying the drawings on each alternate page so far. Complete this table.

| Name | How many faces? | How many edges? | How many vertices? |
|---|---|---|---|
| Cube | 6 | 12 | 8 |
| Rectangular prism | | | |
| Triangular prism | | | |
| Tetrahedron | | | |
| Oblique prism | | | |
| Square pyramid etc. | | | |

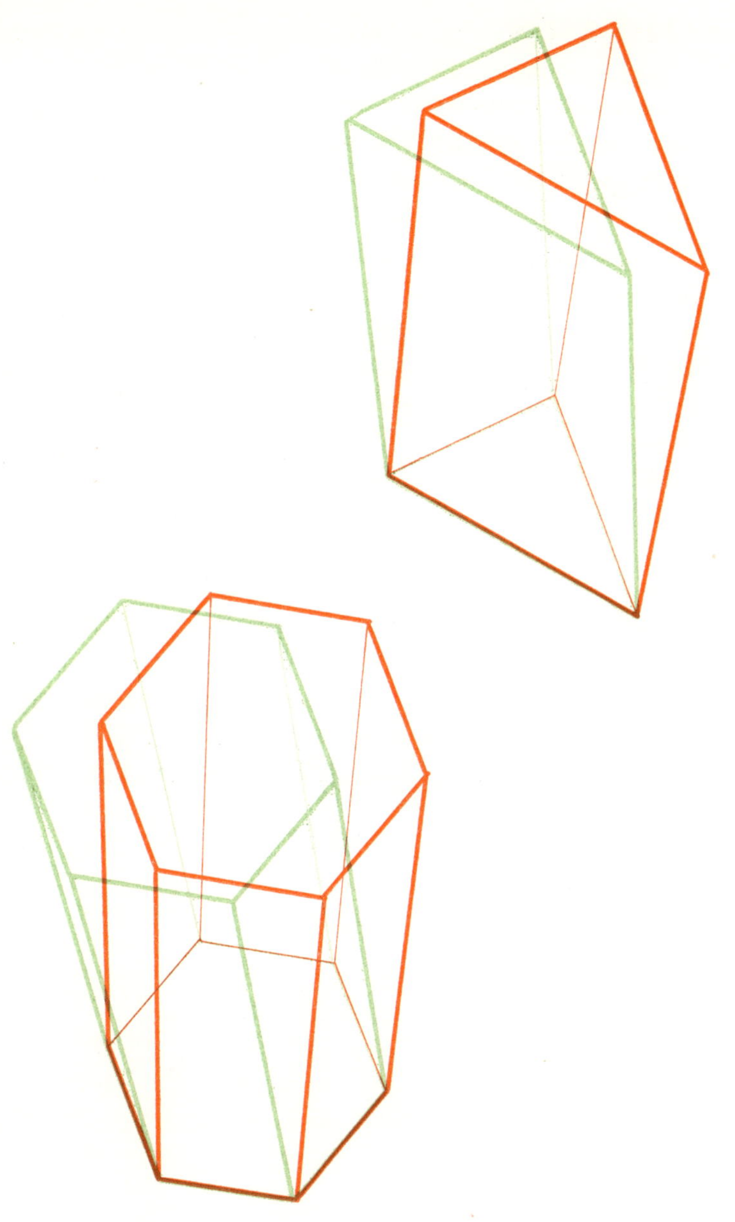

# MORE VOLUMES

You will notice that drawing A on page 17 shows a vertical line which is the vertical height of the square pyramid.

Measure the edge of the base of the pyramid.

Now measure the vertical height. What do you notice?

Drawing B shows the same square pyramid after it has been sheared. Measure a base edge of the sheared pyramid. Is there any change in the lengths of the base edges?

Measure the angle which the base makes with the right-hand edge of the triangular face nearest to you. Where is the vertical height now? Measure it.

We found out earlier that this sort of shearing produced no change in the volume of the solid.

Now examine drawing C. You will notice that three of the sheared pyramids have been fitted together to form a cube. Measure the edges of the cube and compare the lengths with those of the pyramid A. What do you notice? If you know how to find the volume of a cube, you should now be able to find the volume of the pyramid A. What is its volume?

We have used a RIGHT SQUARE PYRAMID for this example in A, that is to say the vertical height runs from the centre of the base to the apex. Its vertical height was the same as the length of a side of the base.

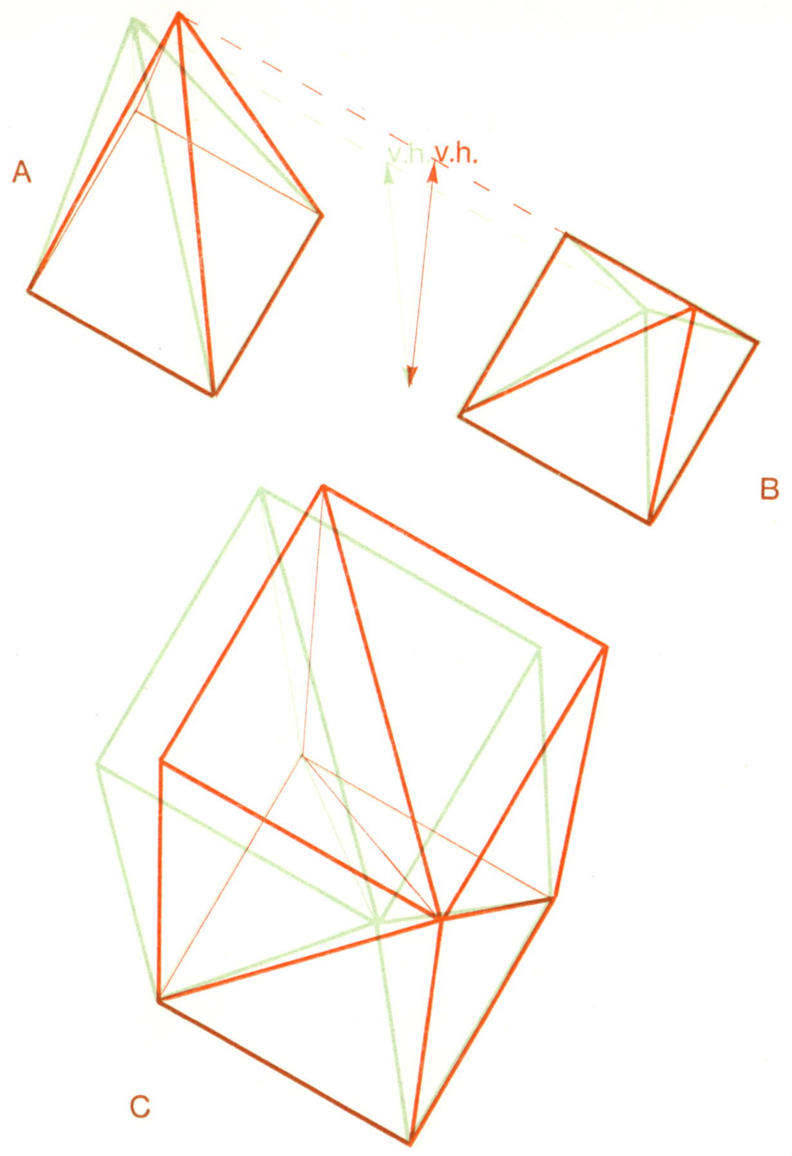
A
v.h. v.h.
B
C

# THE OCTAHEDRON

We have studied two regular solids, the regular tetrahedron and the regular hexahedron or cube.

The drawing A opposite shows the third member of the set of regular solids which is called the REGULAR OCTA-HEDRON.

These solids have names which tell us the number of faces they have, TETRA means four, HEXA means six, OCTA eight.

We also show a net from which a regular octahedron can be constructed. However if you look again at the drawing A you will see that a model of this figure could be made by joining together two right square pyramids by their bases. The slanting edges of the pyramids would be the same length as their base edges.

The second drawing B also shows an octahedron, or solid having eight faces. Examine it and make measurements if you wish. Write down the differences you find between the two shapes.

Would you say that drawing B shows a regular octahedron? Now answer these questions for both figures.
(1) How many faces has an octahedron?
(2) How many edges and vertices are there?
(3) How many faces meet at each vertex?
(4) How many edges meet at each vertex?
(5) What shape are the faces of these figures?
(6) Can you say anything at all about ways of finding the volume of an octahedron?

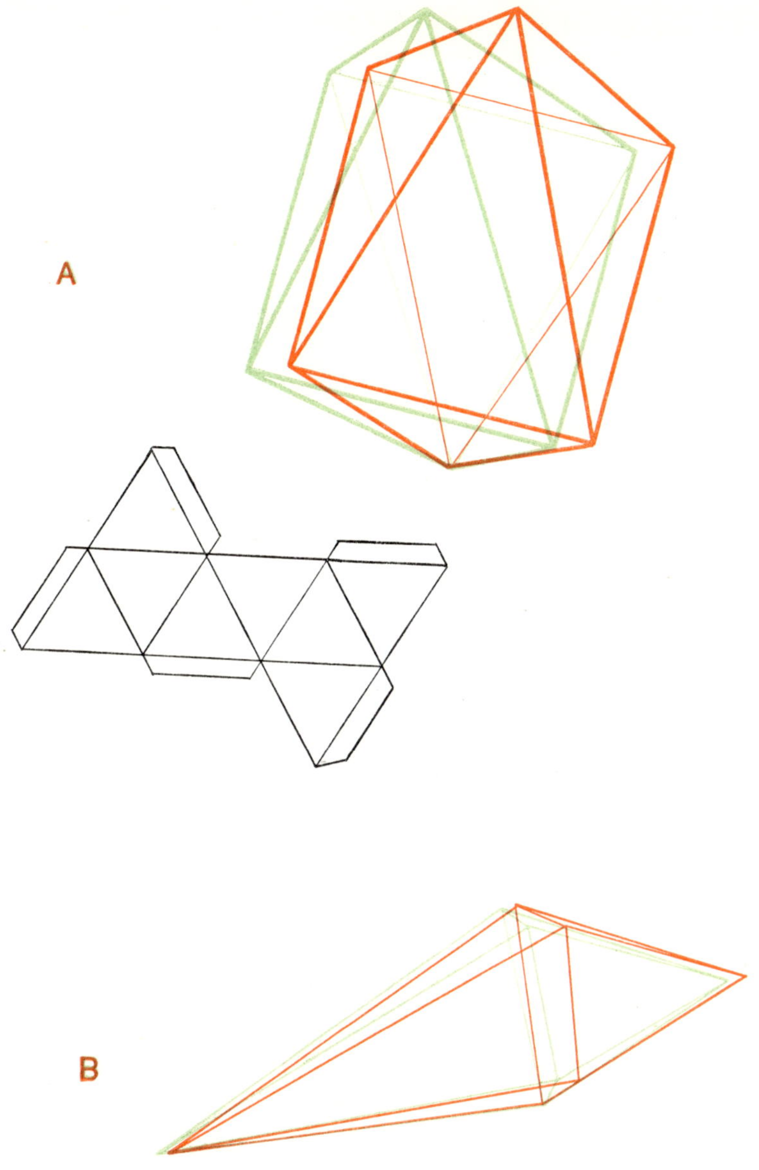
A
B

# MORE REGULAR SOLIDS

THE two regular solids you see on page 21 complete the set. There are only five members of this set. These figures have been known since ancient times.

Drawing A shows the REGULAR DODECAHEDRON and B the REGULAR ICOSAHEDRON.

The names look complicated but you will notice the HEDRON part again and the first part of each name derives from the number of faces.

How many faces has each shape?

What shape are the faces?

How many edges and vertices have they?

How many faces meet at each vertex?

How many edges meet at each vertex?

Measure the edges of each shape.

What do you notice about the lengths of the edges?

Measure an angle where the edges meet.

What size are the angles? Are they all the same size?

How many angles are formed at each vertex?

How could you find the size of the angles at each vertex of a regular solid without measuring, if this is possible?

We have also drawn nets from which these solids can be made.

Try to make the complete set of regular solids.

You may notice that as the number of faces, edges and vertices increases through the set, the shapes gradually appear nearer to the shape of another well known solid. Which one is this?

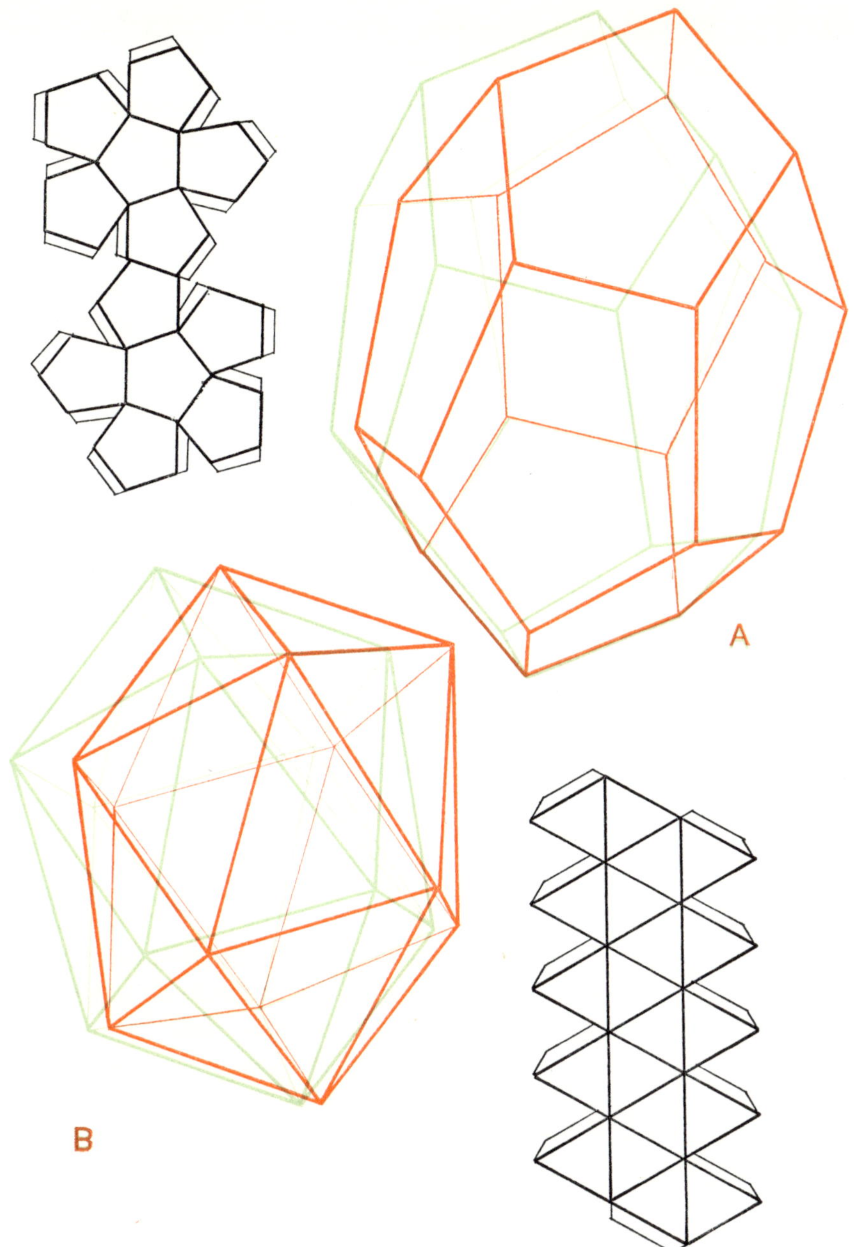
A
B

As we have said, there are five regular solids which were known in ancient times. However in modern times four other solids have been discovered which satisfy most of the conditions for regularity.

There are drawings of two of them, the **GREAT DODECAHEDRON** and the **SMALL STELLATED DODECAHEDRON** on the opposite page. There is also the net of the great dodecahedron from which you may wish to construct this most beautiful shape. Each small rhombus shown is formed from two equilateral triangles.

The dotted lines fold inwards and the solid lines upwards so that the creased edge is towards you.

One point we noticed about the regular solids was that the shape appeared the same no matter which way up the solid was placed.

Look carefully at the two shapes on page 23 and see if the same can be said about them.

In a regular solid the faces are identical.

Is this true of these two shapes?

What shape are the faces?

Are the faces regular plane figures?

In a regular solid the edges are all of the same length.

Is this true of these shapes?

How many edges bound each face?

How many edges meet at each vertex?

A regular solid has vertices which are indistinguishable from one another. What can you say about the vertices in these solids?

How do these shapes differ from a regular solid?

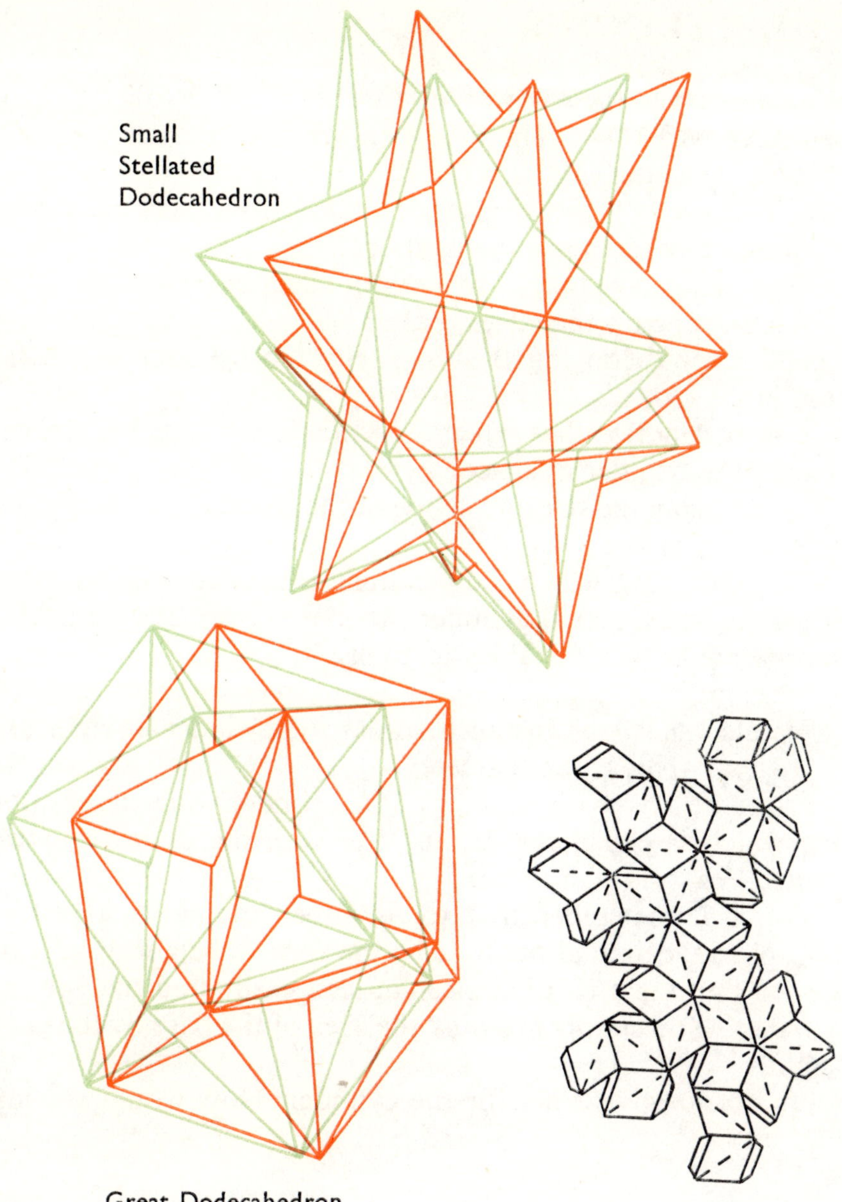
Small
Stellated
Dodecahedron
Great Dodecahedron

# THE CYLINDER

We now turn to the study of another set of solids called SOLIDS OF ROTATION. The first member A of the set you will see on page 25 is the CYLINDER.

It is a sort of prism whose top and bottom faces are circles.

How many faces has the cylinder?

What shape are they? How many edges can you see?

What shape do the edges make?

The second drawing B shows why the cylinder is a solid of rotation.

You will notice that when a rectangle is rotated it traces the shape of a cylinder in space.

By changing the size and shape of the rectangle we can generate different cylinders.

The line about which the rectangle is rotated is called the axis of shape. In an actual cylinder the axis is normally invisible but our special drawing enables us to see it.

Measure its length.

We also show the distance across the cylinder at right angles to the axis. This is the diameter.

From the axis to the surface of the cylinder you will see another line which is called the radius. Measure the diameter and the radius. What do you notice?

You will remember that to find the volume of a prism we multiplied the area of the base by the vertical height. If you know how to find the area of circles you can easily find the volume of cylinders. You simply multiply the area of the base by the vertical height.

We do not give a net for the cylinder. How would you make one?

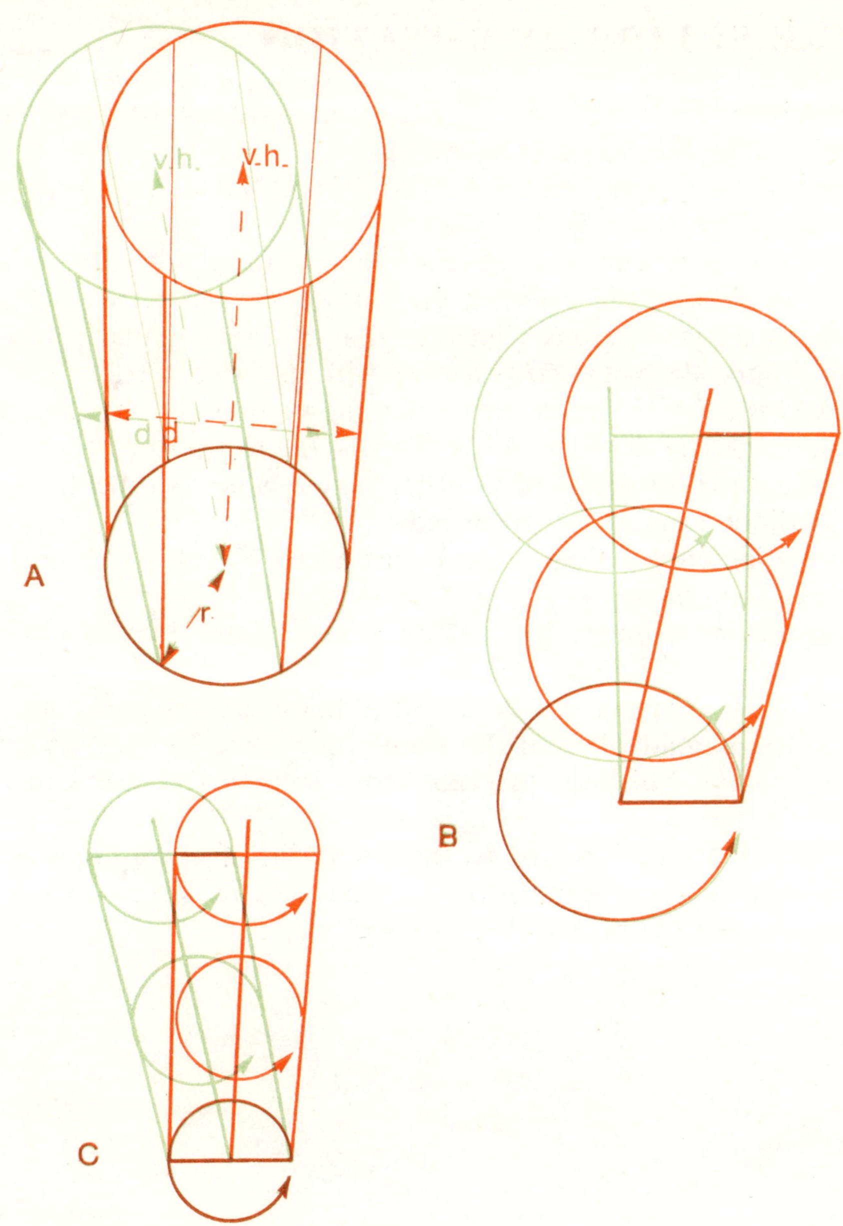

A
v.h. v.h.
d d
r.
B
C

# MORE SOLIDS OF ROTATION

To generate other members of the set of solids of rotation we simply change the shape to be rotated.

The triangle generates a solid called the CONE which is shown opposite. What shape is the base?

The cone is a sort of relative of the pyramid. Its volume is related to that of the cylinder, just as we saw on page 16 the volume of a pyramid was related to that of a rectangular prism.

The second drawing shows a cone with its top cut off.

The shape which traces this solid shows that the top is cut at right angles to the axis. What shape is the top surface?

If we cut the top off a cone so that the cut is not parallel to the base, a different model is produced.

Interesting curved shapes can be produced by cutting through a cone at various angles.

You can experiment by making a cone from Plasticine or clay.

Now draw and cut out some other plane figures and rotate them. You should observe the shape they trace through space as they rotate. This will produce more members of the set of solids of rotation.

If you fix a circular card to the outer rim of a gramophone turntable and allow it to revolve, what sort of shape will be traced in space by the card?

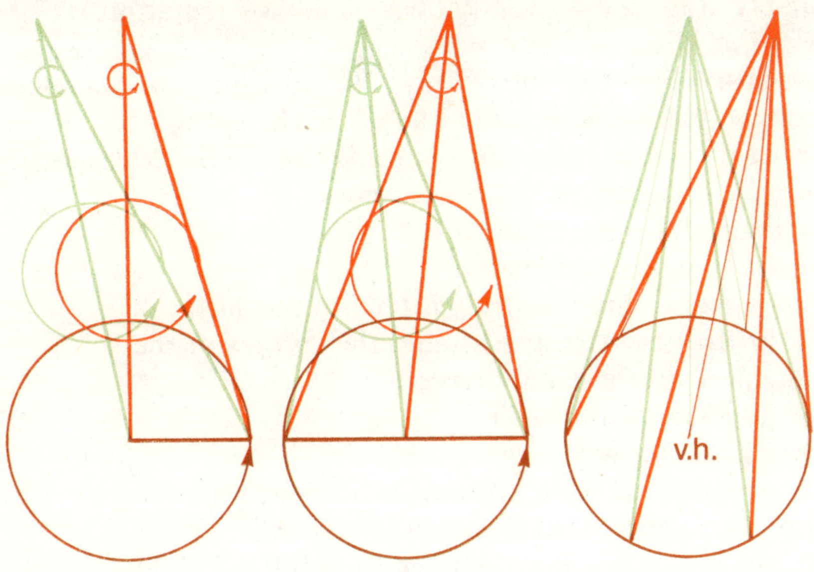
v.h.

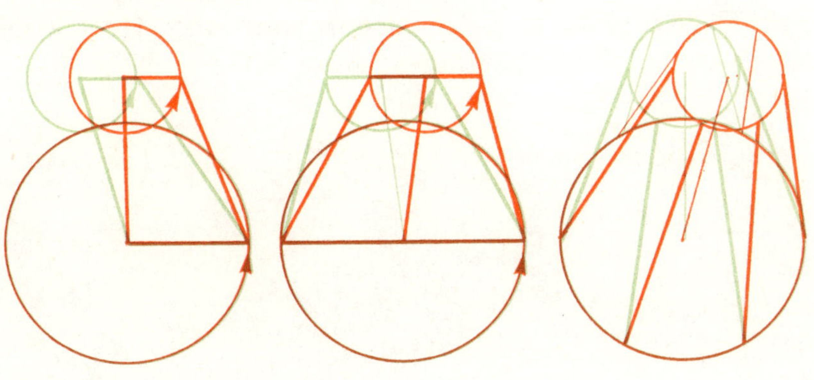

Probably the most well-known solid of rotation is the SPHERE.

Look carefully at the one on page 29 and see if you can decide which shape you could rotate to generate the sphere.

Once again we are able to see inside the shape. Measure the axis and the diameter. What do you find?

How many faces has the sphere?

How many edges and vertices?

Is it possible to draw a straight line on the surface?

Look at the lines running around the surface of the sphere A.

What shape do these lines trace?

Measure some of the radii.

Are they all the same size?

Look at sphere B. Can you see a difference between the lines drawn on this sphere and those on sphere A?

The lines on sphere B are known as GREAT CIRCLES.

Can you say why? Is there a great circle on drawing A?

We have not given a net for the sphere. There are many examples of spheres around you which you can examine.

Next time you peel an orange, try to do it in such a way that you have a net of a sphere. Maybe then you will have some idea of the difficulties faced by people who have to transfer a map of the world from a globe to a flat page in your atlas. How would you do it?

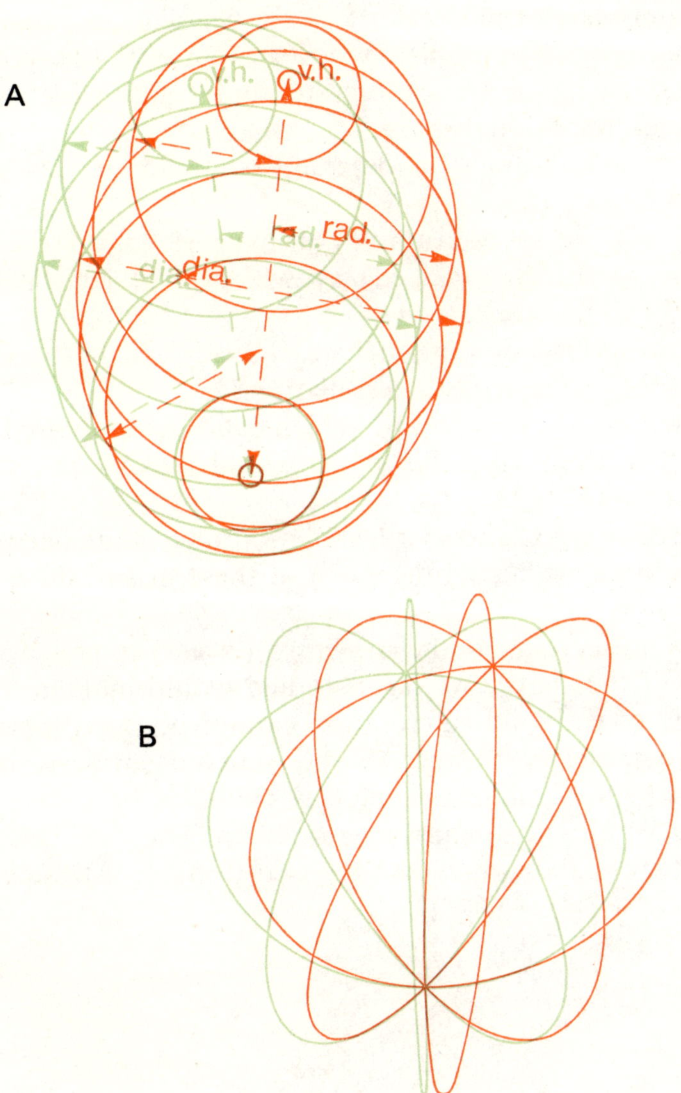

A
v.h.
v.h.
rad.
rad.
dia.
dia.
B

# SOLID TESSELLATIONS

You may have tried making plane tessellations, that is finding flat shapes which completely cover a plane as tiles cover a wall. It is somewhat more difficult to find solids which completely fill three-dimensional space.

We have met one or two examples already in this book.

Which were they?

You will see more examples on page 31.

Look at drawing A and say which solid has been used repeatedly to fill space.

Can you think of any other way in which these shapes could be fitted together to make a tessellation?

Now examine drawing B. Which solid have we used this time?

Do you think that this shape would make a tessellation in a different way?

The drawing C shows a tessellation of rectangular prisms.

Would any other arrangement of these prisms fill space in this way?

Now look back at the drawings on earlier pages and decide which of the shapes we have studied would make a tessellation.

Most of the solids of rotation cannot be fitted together to fill space completely because of their curved surfaces. However if you make some in Plasticine, say cylinders or spheres, you can place several together and squeeze them. They will be compressed into other shapes which can be used to make a tessellation.

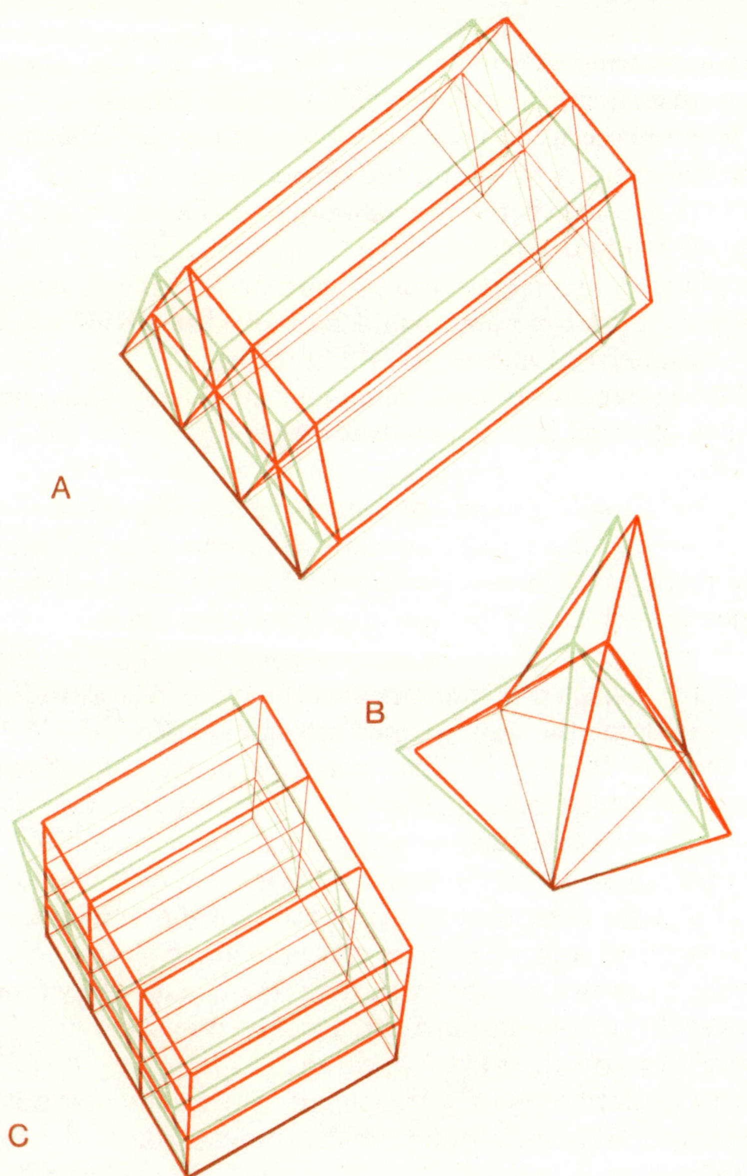

A
B
C

# SYMMETRY

EXAMINE drawing A on page 33. You will see that the cube has been divided into two halves by a plane. What shape is the plane? If the plane were a mirror we would see the other half of the cube reflected in it, giving the appearance of a whole cube. We say that one half is the mirror image of the other half.

When we can find a plane which divides a solid in this way we call it a plane of symmetry. Look again at drawing A and state whether you think the plane could be moved to divide the cube into two symmetrical halves in a different way.

All the regular solids have planes of symmetry along which they can be divided into two halves which are mirror images of each other.

If you have made models of prisms, pyramids and solids of rotation, you can draw lines on the surfaces to show where cuts could be made to form two symmetrical halves. What shapes are the planes in each case after the cuts have been made?

How many planes of symmetry can you find for each solid?

How many planes of symmetry has the cube in drawing A?

Which solid has the most planes of symmetry?

Now look at drawing B. You will see that we have given the cube an axis, which passes from the centre of one face to the centre of the opposite face. If we spun a cube around such an axis it could come to rest in four positions where it would look exactly the same even though we were viewing different faces. This axis is called a fourfold axis of symmetry. Why?

Drawing C shows a cube with a different axis of symmetry. If we rotated the cube around this axis, in how many positions could it rest to give the same appearance?

You can explore axes of symmetry in other solids by pushing a needle through them, around which you can rotate the figure. If you number each face of the solid it will help you to check its various positions as you rotate it.

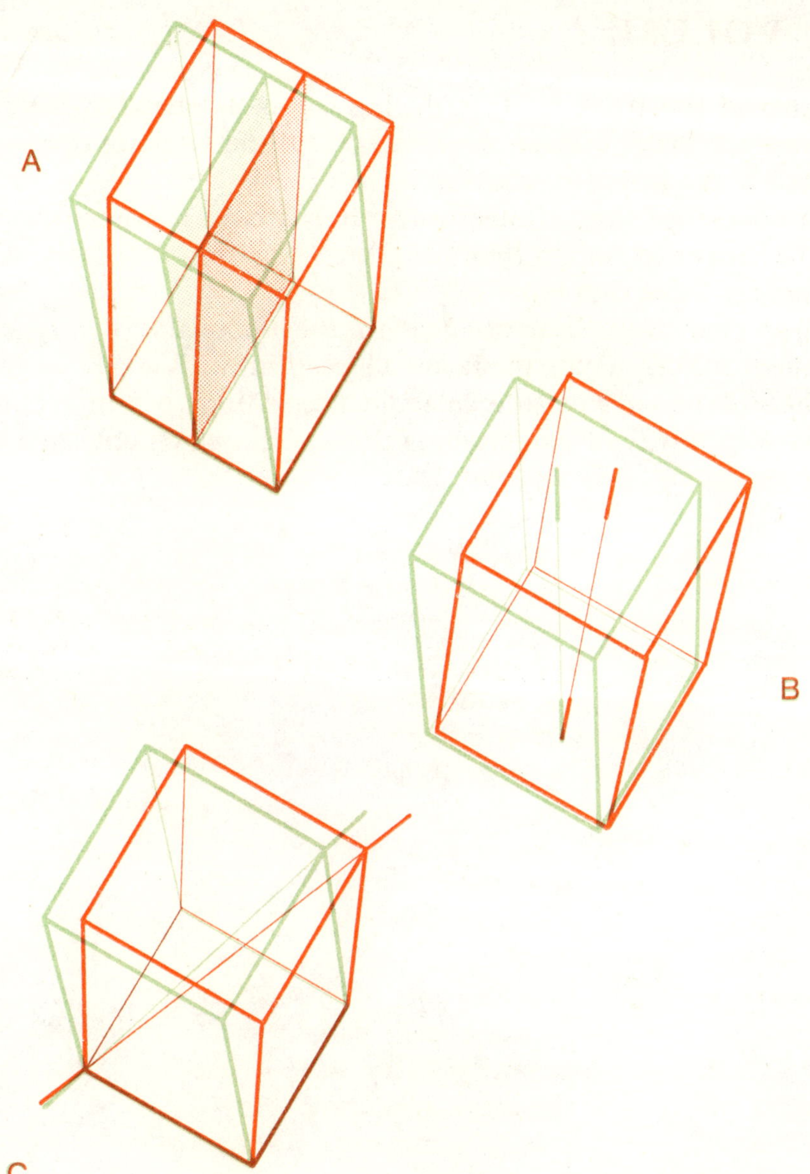
A
B
C

# NO VOLUME?

We began the book by introducing solids as shapes which fill a space and having three dimensions. A little later we discussed volume and ways of measuring it.

Of course all three-dimensional shapes have a volume which can be measured, or do they?

Look at these shapes.

They bear some resemblance to the cube and the regular tetrahedron, but with each face collapsed inwards.

These shapes also make splendid decorations if you are able to think of a way of constructing them, with which challenge we leave you.

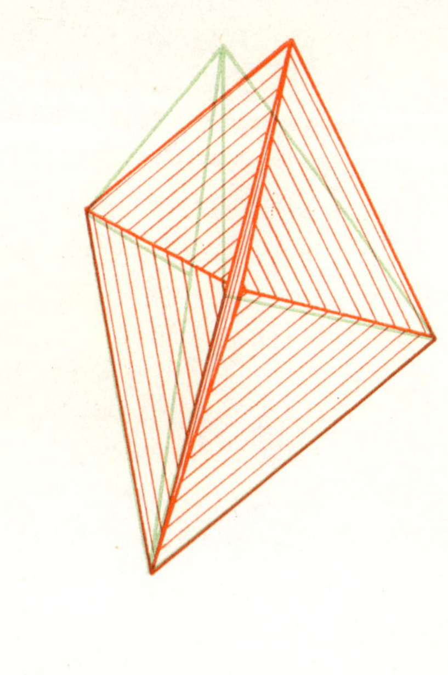

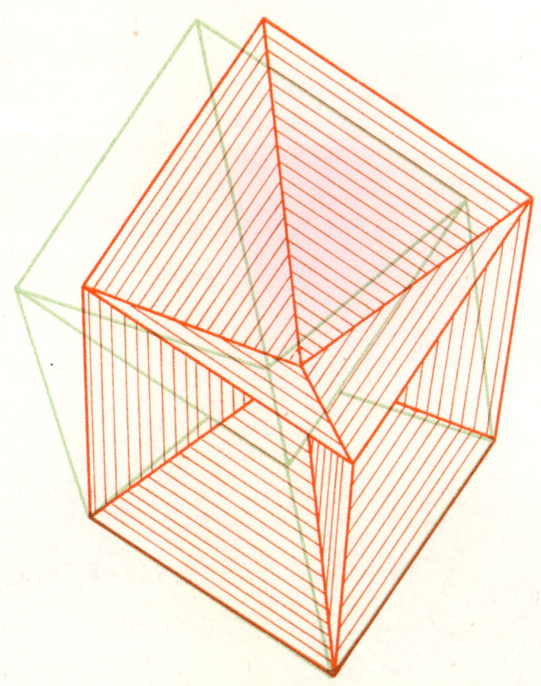